ENERGY FROM MOVING WATER

by Karen Latchana Kenney

Consultant: Beth Gambro
Reading Specialist, Yorkville, Illinois

BEARPORT
PUBLISHING

Minneapolis, Minnesota

Teaching Tips

Before Reading

- Look at the cover of the book. Discuss the picture and the title.

- Ask readers to brainstorm a list of what they already know about water and energy. What can they expect to see in this book?

- Go on a picture walk, looking through the pictures to discuss vocabulary and make predictions about the text.

During Reading

- Read for purpose. Encourage readers to think about water and energy and the roles they play in our daily lives as they are reading.

- Ask readers to look for the details of the book. What are they learning about water?

- If readers encounter an unknown word, ask them to look at the sounds in the word. Then, ask them to look at the rest of the page. Are there any clues to help them understand?

After Reading

- Encourage readers to pick a buddy and reread the book together.

- Ask readers to name one reason to use dams and one reason to not use dams. Go back and find the pages that tell about these things.

- Ask readers to write or draw something they learned about energy from water.

Credits:
Cover and title page, © Gary Saxe/Shutterstock; 3, © CK Foto/Shutterstock; 4 –5, © Anatoliy Karlyuk/Shutterstock; 6–7, © ilkercelik/iStock, © CK Foto/Shutterstock; 8–9, © 4FR/iStock; 10–11, © Claudiad/iStock; 12, © Satakorn/Shutterstock; 13, © atanarus/Shutterstock; 15, © James Pintar/iStock; 16–17, © Bim/iStock; 18–19, © David McNew/Getty; 20, © Reimphoto/iStock; 22, © bubblea/iStock; 23BL, © Mumemories/iStock; 23BR, © Andrey_Popov/Shutterstock; 23TL, © Mikhail Leonov/Shuterstock; 23TM, © ventdusud/iStock; 23TR, © Prykhodov/iStock

Library of Congress Cataloging-in-Publication Data

Names: Kenney, Karen Latchana, author.
Title: Energy from moving water / by Karen Latchana Kenney.
Description: Minneapolis, Minnesota : Bearport Publishing Company, [2022] |
Series: Power up with energy! | "Bearcub books." | Includes bibliographical references and index.
Identifiers: LCCN 2020054974 (print) | LCCN 2020054975 (ebook) | ISBN
9781647478650 (library binding) | ISBN 9781647478728 (paperback) | ISBN
9781647478797 (ebook)
Subjects: LCSH: Hydroelectric power plants--Juvenile literature.
Classification: LCC TK1081 .K45 2022 (print) | LCC TK1081 (ebook) | DDC
621.31/2134--dc23
LC record available at https://lccn.loc.gov/2020054974
LC ebook record available at https://lccn.loc.gov/2020054975

For more information, write to Bearport Publishing,
5357 Penn Avenue South, Minneapolis, MN 55419.
Printed in the United States of America.

Contents

Powering the Tablet

Oh, no!

The **tablet** is dead.

How can we make it work again?

We can use moving water!

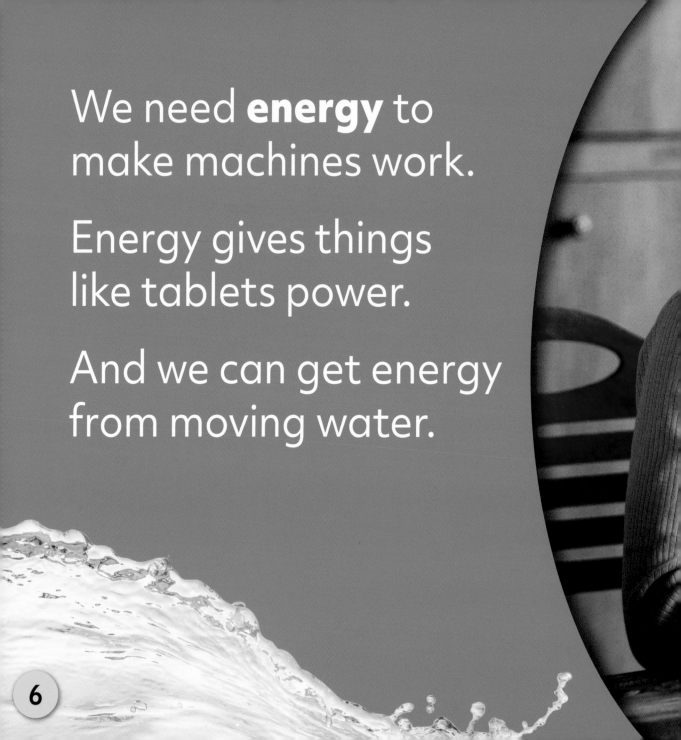

We need **energy** to make machines work.

Energy gives things like tablets power.

And we can get energy from moving water.

Big rivers have lots of water.

The water moves fast.

It has energy we can use.

A power **plant** can get the energy.

The plant is built in a wall that stops the river's water.

The wall is called a **dam**.

River water moves into the dam.

The water's moving energy turns the **blades**.

This makes power.

A blade

Then, the power travels in wires to our homes.

Machines use that power.

Our TVs work with the power.

We can use it to turn on lights, too.

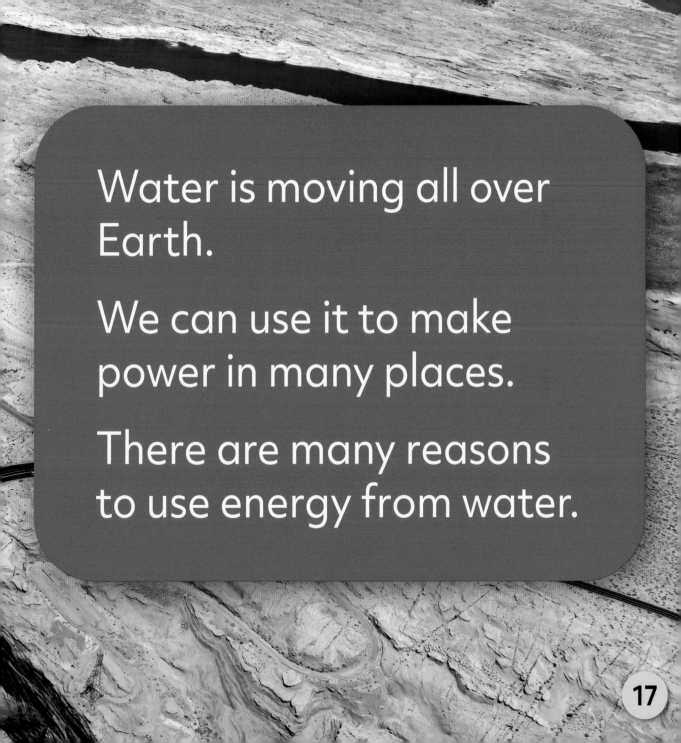

Water is moving all over Earth.

We can use it to make power in many places.

There are many reasons to use energy from water.

But dams can slow rivers.

People and animals may be hurt by this.

Dams may move water away from plants that need it, too.

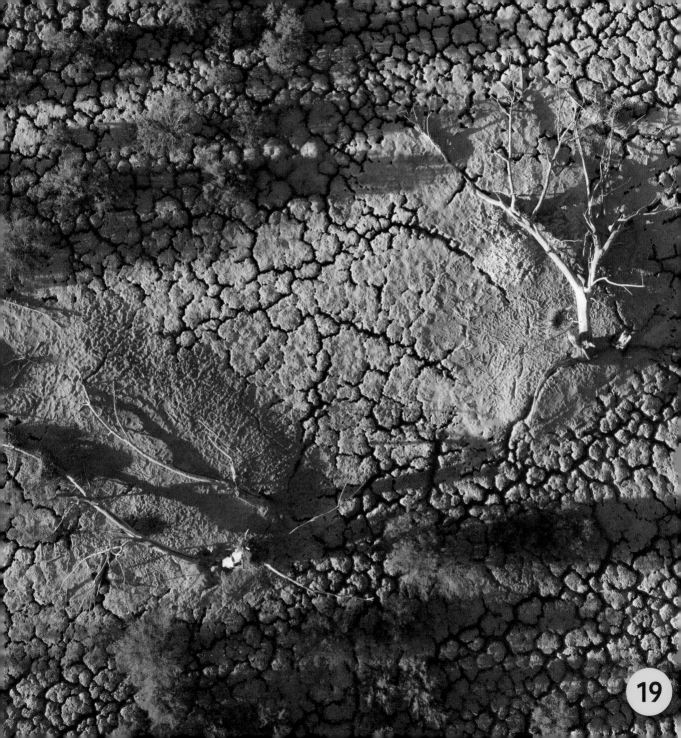

People are trying to be smarter with water energy.

They want to make dams better for everyone.

Soon, moving water will make more of the power we need.

Energy from Moving Water

Follow along as moving water makes power.

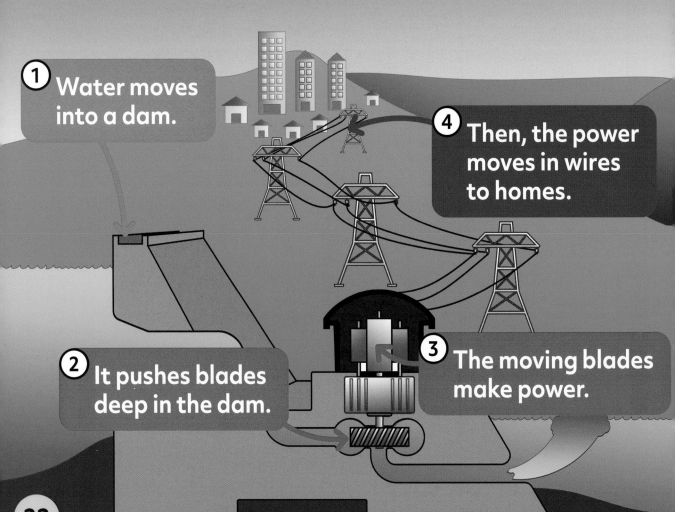

1. Water moves into a dam.

2. It pushes blades deep in the dam.

3. The moving blades make power.

4. Then, the power moves in wires to homes.

Glossary

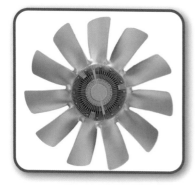

blades flat spinning parts that are used on some machines

dam a large wall that holds back water

energy power that makes things work

plant a place where power is made and gathered

tablet a small computer that is easy to carry around

Index

Read More

Bowman, Chris. *Dams (Blastoff! Readers: Everyday Engineering)*. Minneapolis: Bellwether, 2019.

Olson, Elsie. *Water Energy (Earth's Energy Resources)*. Minneapolis: Abdo, 2019.

Learn More Online

1. Go to **www.factsurfer.com**
2. Enter "**Water Energy**" into the search box.
3. Click on the cover of this book to see a list of websites.

About the Author

Karen Latchana Kenney likes biking and reading. She tries to find ways to use less energy every day.